YOUR KNOWLEDGE HAS VALUE

Bibliographic information published by the German National Library:

The German National Library lists this publication in the National Bibliography; detailed bibliographic data are available on the Internet at http://dnb.dnb.de .

Imprint:

Copyright © 2016 GRIN Verlag, Open Publishing GmbH
Print and binding: Books on Demand GmbH, Norderstedt Germany
ISBN: 9783668365094

This book at GRIN:

http://www.grin.com/en/e-book/347012/drug-resistance-in-mycobacterium-tuberculosis

Tai Man Chan

Drug resistance in Mycobacterium Tuberculosis

GRIN Publishing

GRIN - Your knowledge has value

Since its foundation in 1998, GRIN has specialized in publishing academic texts by students, college teachers and other academics as e-book and printed book. The website www.grin.com is an ideal platform for presenting term papers, final papers, scientific essays, dissertations and specialist books.

Visit us on the internet:

http://www.grin.com/

http://www.facebook.com/grincom

http://www.twitter.com/grin_com

Drug resistance in *Mycobacterium Tuberculosis*

Author: Chan Tai Man

Abstract

Tuberculosis (TB) has been one of the most serious public health problems in the world. Although many years of hard work by World Health Organization and different countries over the tuberculosis control programs, TB remains one of the top 10 causes of death worldwide (World Health Organization 2016a). The control programs are greatly hampered with the emergence of multidrug resistance tuberculosis (MDR-TB) and severe forms of drug resistance like extensively drug resistance tuberculosis (XDR-TB). In 2015, 10.4 million people were infected with tuberculosis while 1.8 million people were killed (World Health Organization 2016a). In this review, the drugs used to treat tuberculosis, which is classified into the first line and second line drugs, the mechanisms of drugs resistance and the diagnosis of tuberculosis will be discussed.

Content

Introduction

Tuberculosis is caused by bacteria *Mycobacterium tuberculosis* that is in the family Mycobacteriaceae (Ryan et al. 2004). It was first discovered by Robert Koch in 1882. *M. tuberculosis* has an unusual, waxy coating on its cell surface, which is attributed to the presence of mycolic acid. This makes the cells impervious to Gram staining. In clinical settings, it can be appeared in the form of Gram negative and Gram positive (Fu et al. 2000).

After the introduce of first anti-TB drug, streptomycin (STR), para-aminosalicylic acid (PAS), isoniazid (INH), drug resistance to these drugs were found in clinical isolates of *M. tuberculosis* (Crofton and Mitchison 1948). In order to cure the illness, physicians have to measure the drug resistance accurately and promptly. A critical proportion method was introduced by the Pasteur Institute in 1961 for drug susceptibility testing in TB. Eventually, this becomes a standard method for measuring drug resistance (Espinal 2003). Studies in various countries in the 1960s have shown that drug resistance in developing countries is higher than that of developed countries (Espinal 2003). After the usage of effective anti-TB drug rifampicin (RIF) in the 1960s and with the use of combination therapy, there was a drop in drug resistant and drug susceptible TB in the developed countries. This leads to a drop in interest and, as a result, funding in TB control programs. So, no concrete study of drug resistance on TB was performed and recorded for the following 20 years (Espinal 2003). In addition, HIV / AIDS in the 1980s led to an increase in TB transmission, which is also associated to outbreaks of multi drug resistant TB (MDR-TB), which is the resistant to RIF and INF (Edlin et al. 1992, Fischl et al. 1992).

Although TB is curable and preventable in most cases, a lack of control in TB management would lead to wide spread of TB and MDR-TB worldwide. Around 17% of the infected patients were killed in 2015 (World Health Organization 2016b). In 2015, there were an estimated number of 480,000 people (4.6% of total infected people) were suffered from multidrug-resistant TB (MDR-TB) (World Health Organization 2016b). With the high death rates assiociated with MDR TB of around 50% to 80% and spans a relatively short time of around 4 to 16 weeks from diagnosis to death (Mathew et al. 2014). As a result, the Sustainable Development Goals (SDGs) and the End TB Strategy were introduced in 2015, which covers the period from 2016-2030 and 2016-2035 respectively (World Health Organization 2016b). Among all cases, the situation in the developing countries is invariably worse than the developed countries as high as 95% of the TB deaths occur in the low and middle income countries. 6 countries account for 60% of the cases in descending order: India, Indonesia, China, Nigeria, Pakistan and South Africa. In most cases, the TB drug resistance emerges when the anti-TB medicines are used inappropriately, incorrect prescription s given by health care providers, poor quality drugs and termination of treatments prematurely by patients from various reasons (World Health Organization 2016a).

Molecular mechanism of drug resistance

The mechanism of how *M. tuberculosis* develops drug resistances has to be investigated in order to better control the spreading of tuberculosis in the world. This also helps to prevent or control the development of drug resistance tuberculosis and to identify genes associated with drug resistance of new drugs (Johnson et al. 2006). Genetic and molecular analysis of *M. tuberculosis* suggests that accumulation of mutations in the drug target genes is the primarily contribution factor to drug resistance in *M. tuberculosis*. The mutations either lead to an altered target, for example, RNA polymerase and catalase-peroxidase in rifampicin and isoniazid resistance respectively, or a change in titration of the drug, like InhA inisoniazid resistance (Rattan et al. 1998). Resistance to anti-TB drugs can occur spontaneously with an estimated frequency of 3.5×10^{-6} for INH and 3.1×10^{-8} for RIF when there are mutations in the genome of *M. tuberculosis* (Johnson et al. 2006). As the chromosomal loci contributing to the resistance of different are not linked, the probability of drug resistance to both INH and RIF are usually extremely low, as the chance is the multiplication of the 2 independent frequencies, which is 9×10^{-14} (Dooley and Simone 1994). Apart from mutations of the genome of *M. tuberculosis*, the other occurrence of drug resistance is the acquired resistance. This arises when drug resistant mutants are selected as a result of ineffective treatment or patient non-compliance or as primary resistance when a patient is infected with a resistant strain (Johnson et al. 2006).

To effectively control the wide spread of TB across the world, (World Health Organization 2016b) has recently issued a new guidance in May 2016 that "all cases of rifampicin-resistant TB (RR-TB, with or without resistance to other drugs), including those with multidrug-resistant TB (MDR-TB), should be treated with a second-line MDR-TB treatment regimen. Globally in 2015, there were an estimated 480 000 new cases of MDR-TB and an additional 100 000 people with rifampicin-resistant TB who were also newly eligible for MDR-TB treatment." Treatment can be divided into 2 categories of drugs: First line drugs and second line drugs.

First line drugs used in TB treatment

In the anti-TB regiment, the drugs should have an effective sterilizing activity which is effective in shortening the duration of the treatment. (Johnson et al. 2006). There are 4 first line drugs in treating TB are ethambutol (EMB), isoniazid (INH), pyrazinamide (PZA) and rifampicin (RMP) (Johnson et al. 2006). Rifampicin and isoniazid are considered to be the two most effective and powerful anti-TB drugs (World Health Organization 2016b). In most cases, a four-drug regiment is used for the TB treatment (Johnson et al. 2006).

Resistance to the first line anti-TB drugs are linked to the mutations of at least 10 genes: *katG, inhA, ahpC, kasA* and *ndh* for INH resistance; *rpoB* for RIF resistance, *embB* for EMB resistance, *pncA* for PZA resistance and *rpsL* and *rrs* for STR resistance (Johnson et al. 2006). The drug resistance for the 4 major first line drugs will be discussed below: INH, RIF, EMB and PZA. Isoniazid, or isonicotinic acid hydrazide (INH), was first synthesized in the early

19th century. However, the ability on treating TB was first known in 1951 (Barry Iii et al. 1998, Rattan et al. 1998, Slayden and Barry Iii 2000).

katG. INH enters the cell as a prodrug which is then converted by a catalase peroxidase encoded by *katG.* The peroxidase activity of the enzyme enables to activate INH to a toxic substance in the bacterial cell (D. Xu et al. 1996). The toxic substance then affects intracellular targets like mycolic acid biosynthesis which are an important component of the M. TB cell wall. The deterioration of mycolic acid synthesis gradually results in loss of cellular integrity and the bacteria are killed (Barry Iii et al. 1998). Middlebrook (1954) had demonstrated that a loss of catalase activity leads to INH resistance (Middlebrook 1954). The genetic studies found that transformation of INH-resistant *Mycobacterium smegmatis* and *M. tuberculosis* strains with a functional *katG* gene restored INH susceptibility and *katG* deletions give rise to INH resistance (Heym et al. 1999, Zhang et al. 1993, Zhang et al. 1992).

ahpC. Very often, a loss of *katG* activity due to the S315T amino acid substitution is accompanied by an increase in expression of *ahpC,* an alkyl hydroperoxide reductase protein which is capable of detoxifying damaging organic peroxides(Sherman et al. 1996). It was found that 5 different nucleotide alterations in the promoter region of the *ahpC* gene would lead to over expression of *ahpC* and subsequently INH resistance (Ramana Rao et al. 2011). The overexpression gives a detoxifying effect on organic peroxides within the cell and protects the bacteria against oxidative damage but not for INH. (Johnson et al. 2006).

inhA. The protein encoded by the *inhA* locus is one of the targets for activated INH. *InhA* is an enoyl-acyl carrier protein (ACP) reductase that is proposed to be the primary target for drug resistance to INH and one second line drug, ethionamide (ETH) (Banerjee et al. 1994). A ternary complex is formed when activated INH binds to the *InhA*-NADH complex that results in inhibition of mycolic acid biosynthesis. Six point mutations are associated with INH resistance within the structural *inhA* gene have been identified (Basso and Blanchard 1998, Ramaswamy et al. 2000).

kasA. There are considerable dispute on whether *kasA* is a possible target for INH resistance (Sherman et al. 1996). *kasA* encodes a â-ketoacyl-ACP synthase which is to synthesize mycolic acids. The mutations of this gene usually confer a low level of INH resistance. Genotypic analysis of the *kasA* gene shows 4 different amino acid substitutions involving codon 66 (GAT-AAT), codon 269 (GGT-AGT), codon 312 (GGC-AGC) and codon 413 (TTC-TTA) (Mdluli et al. 1998, Ramaswamy et al. 2000).

Ndh. Miesel et al. (1998) found a new INH resistance in M. semgmatis in 1998. The ndh gene encodes NADH dehydrogenase which is bound to the active site of *inhA* and a ternary complex with activated INH is formed. Structural studies have found that a reactive form of INH attacks the NAD(H) co-factor and a covalent INH-NAD adduct is generated. Mutations in the *ndh* gene lead to defects in the enzymatic activity. As a result, defects in the oxidation of NADH to NAD result in NADH accumulation and NAD depletion (Lee et al. 2001). The accumulation of NADH contributes to the inhibition of the binding effects of the INH-NAD adduct to the active site of the *InhA* enzyme (Miesel et al. 1998, Rozwarski et al. 1998). Rifampicin was introduced in 1972 as an anti-TB drug and together with INH are the two most effective first line drugs in short-course chemotherapy (Ramaswamy and Musser 1998, Ramaswamy et al. 2000). The action of RIF in combination with PZA shortens the routine TB treatment from 1 year to 6 months.

rpoB. The action of RIF is to interfere with transcription with the DNA-dependent RNA polymerase. It is composed of four different subunits (á, â, â' and ó) which are encoded by *rpoA, rpoB, rpoC* and *rpoD* genes respectively. RIF kills the organism by binding to the â-subunit hindering transcription. A number of mutations and short deletions in the *rpoB* gene were identified by extensive studies on the gene in RIF resistant isolates of *M. tuberculosis* (Johnson et al. 2006). A number of 69 single nucleotide changes, 3 insertions, 16 deletion and 38 multiple nucleotide changes have been reported (Herrera et al. 2013). These changes were found in more than 70% of RIF resistant isolates (Herrera et al. 2013, Ramaswamy and

Musser 1998, Ramaswamy et al. 2000). In addition, minimal inhibitory concentration showed mutations in coden 526 and 531 contributes to a high level of RIF resistance (Johnson et al. 2006).

Pyrazinamide (PZA) is a nicotinamide analog. The anti-TB activity was first discovered in 1952. PZA targets a fatty-acid synthesis enzyme and is in charge of killing persistent tubercle bacilli in the initial intensive phase of chemotherapy (Somoskovi et al. 2001). In addition, PZA has effective sterilizing activity that effectively shortens the chemotherapeutic regiment from 12 to 6 months. PZA, similar to INH, is a prodrug and is converted to the active form, pyrazinoic acid (POA) by the pyrazinamidase (PZase) encoded by *pncA* (Johnson et al. 2006). The activity of PZA is highly specific for *M. tuberculosis* only, while there is no effect on other mycobacteria (Johnson et al. 2006). PZA is only active in an acidic medium where POA accumulates in the cytoplasm due to an ineffective efflux pump. The accumulation results in a lowering of intracellular pH to a level that inactivates a vital fatty acid synthase (Zimhony et al. 2004). Cloning and characterization of the *M. tuberculosis pncA* gene showed that *pncA* mutations lead to PZA resistance (Scorpio and Zhang 1996). Different kinds of *pncA* mutations have been identified in more than 70% of PZA resistant clinical isolates scattered throughout the *pncA* gene (Scorpio et al. 1997, Scorpio and Zhang 1996, Sreevatsan et al. 1997a).

Ethambutol (EMB) is used in combination with other drugs when treating TB and is specific to the mycobacteria. EMB inhibits an arabinosyl transferase (*embB*) involved in cell wall iosynthesis (Takayama and Kilburn 1989). Studies by Telenti et al. found out 3 genes, designated embCAB, that encode homologous arabinosyl transferase enzymes involved in EMB resistance (Telenti et al. 1997). Further studies were performed and five mutations in codon 306 [(ATG-GTG), (ATG-CTG), (ATG-ATA), (ATG-ATC) and (ATG-ATT)] which

8

result in three different amino acid substitutions (Val, Leu and Ile) in EMB-resistant isolates were found (Lee et al. 2002, Mokrousov et al. 2002, Sreevatsan et al. 1997b). These five mutations associated with a high rate of 70–90% of all EMB resistant isolates (Ramaswamy et al. 2000).

Second line drugs used in TB treatment

When treating the MDR-TB, which does not respond to the most powerful first line drug, isoniazid and rifampicin, second line drugs are used (World Health Organization 2016a). But the treatment options are limited and require extensive chemotherapy which usually takes up to 2 years of treatment and the medicines are also expensive and toxic (World Health Organization 2016a).

The following drugs are classified as second line drugs and will be discussed: aminoglycosides (kanamycin and amikacin) polypeptides (capreomycin, viomycin and enviomycin), fluoroquinolones (ofloxacin, ciprofloxacin, and gatifloxacin), D-cycloserine and thionamides (ethionamide and prothionamide) (World Health Organization 2001).

Ciproflaxin (CIP) and Ofloxacin (OFL) are the two fluoroquinolones (FQs) used in MDR-TB treatment (World Health Organization, 2001). The quinolones inactivate DNA gyrase, a type II DNA topoisomerase (Cynamon and Sklaney 2003, Ginsburg et al. 2003, Rattan et al. 1998). DNA gyrase is encoded by *gyrA* and *gyrB* (Takiff et al. 1994). The negative supercoils in closed circular DNA molecules are introduced (Ramaswamy and Musser 1998, Rattan et al. 1998). The quinolone resistance-determining region (QRDR) is a conserved region in the *gyrA* (320bp) and *gyrB* (375bp) genes (Ginsburg et al. 2003). These 2 genes are interaction points of FQ and gyrase (Ginsburg et al. 2003). Missense mutations in codon 90, 91, and 94 of *gyrA* gene are the main factors for resistance to FQs (Takiff et al. 1994, Chen Xu et al. 1996)

.

Kanamycin (KAN) and Aminokacin (AMI) are aminoglycosides that inhibit protein synthesis and so is not able to treat dormant *M. tuberculosis* (Johnson et al. 2006). Aminoglycosides function is to bind to bacterial ribosomes and disturb the elongation of the peptide chain in the bacteria (Johnson et al. 2006). Mutations in the *rrs* gene encoding for 16s rRNA are associated with drug resistance to KAN and AMI (Suzuki et al. 1998).

Ethionamide (ETH) is mechanistically and structurally similar to INH. ETH is also a prodrug activated by bacterial metabolism. The activated drug then has the ability to disrupt cell wall biosynthesis by inhibiting mycolic acid synthesis. Similar to INH, mutations in the promoter of the *inhA* gene are associated with drug resistance to ETH (Morlock et al. 2003).

D-Cycloserine (DCS), a cyclic analog of D-alanine, which is one of the central molecules of the cross linking step of peptidoglycan assembly (Cáceres et al. 1997, David 2001, Feng and Barletta 2003, Ramaswamy et al. 2000). DCS inhibits cell wall synthesis through competing with D-Alanine for two enzymes D-alanyl-D-alanine synthetase (Ddl) and D-alanine racemase (Alr) and inhibiting the synthesis of the enzymes proteins. An over expression of alr leads to drug resistance (Feng and Barletta 2003, Ramaswamy and Musser 1998).

Diagnosis

Molecular methods are often used for the identification of mutations in resistance-causing genes provides a fast means to screen *M. tuberculosis* isolates for antibiotic resistance (Johnson et al. 2006). Mutation screening methods includes: DNA sequencing, probe based hybridization methods, PCR-RFLP, single-strand conformation polymorphism (SSCP), heteroduplex analysis (HA), molecular beacons and ARMS-PCR (Victor et al. 2002).

The use of the rapid test *Xpert* MTB/RIF® has expanded substantially since 2010, when WHO first recommended its use (World Health Organization 2016b). This method was developed by Cepheid, USA. The test simultaneously detects TB and resistance to rifampicin.

Diagnosis can be made within 2 hours and is recommended by WHO as the initial diagnostic test in all persons with signs and symptoms of TB (World Health Organization 2016b). In 2015, more than 100 countries are using *Xpert* and 6.2 million cartridges were procured globally. In 2016, 4 new diagnostic tests were recommended by WHO – a rapid molecular test to detect TB at peripheral health centres where *Xpert* MTB/RIF cannot be used, and 3 tests, microscopy, WRDs (other than *Xpert* MTB/RIF), culture and drug susceptibility testing (DST), to detect resistance to first- and second-line TB medicines (World Health Organization 2016b)

Conclusion

In conclusion, tuberculosis threatens millions of lives per year on earth. Recently, the number of MDR-TB cases remains very high and as the death rate of MDR-TB is significantly higher than TB, the WHO has launched different TB campaigns. The molecular perspective mechanisms of how drug resistant is developed in first line and second line TB drugs are reviewed and the diagnoses methods used in 1990s and the recent diagnoses methods recommended by WHO are also discussed.

(2,980 words)

Reference list

Banerjee, A., Dubnau, E., Quemard, A., Balasubramanian, V., Um, K., Wilson, T., Collins, D., de Lisle, G. and Jacobs, W. (1994) inhA, a gene encoding a target for isoniazid and ethionamide in Mycobacterium tuberculosis. *Science, 263*(5144), pp. 227-230.

Barry Iii, C. E., Lee, R. E., Mdluli, K., Sampson, A. E., Schroeder, B. G., Slayden, R. A. and Yuan, Y. (1998) Mycolic acids: structure, biosynthesis and physiological functions. *Progress in Lipid Research, 37*(2–3), pp. 143-179.

Basso, L. A. and Blanchard, J. S. (1998) Resistance to Antitubercular Drugs. in Rosen, B. P. and Mobashery, S., (eds.) *Resolving the Antibiotic Paradox: Progress in Understanding Drug Resistance and Development of New Antibiotics*,Boston, MA: Springer US. pp. 115-144.

Cáceres, N. E., Harris, N. B., Wellehan, J. F., Feng, Z., Kapur, V. and Barletta, R. G. (1997) Overexpression of the D-alanine racemase gene confers resistance to D-cycloserine in Mycobacterium smegmatis. *Journal of Bacteriology, 179*(16), pp. 5046-5055.

Crofton, J. and Mitchison, D. A. (1948) Streptomycin Resistance in Pulmonary Tuberculosis. *British Medical Journal, 2*(4588), pp. 1009-1015.

Cynamon, M. H. and Sklaney, M. (2003) Gatifloxacin and Ethionamide as the Foundation for Therapy of Tuberculosis. *Antimicrobial Agents and Chemotherapy, 47*(8), pp. 2442-2444.

David, S. (2001) Synergic activity of d-cycloserine and â-chloro-d-alanine against

Mycobacterium tuberculosis. *Journal of Antimicrobial Chemotherapy, 47*(2), pp. 203-206.

Dooley, S. W. and Simone, P. M. (1994) The Extent and Management of Drug-Resistant Tuberculosis: The American Experience. *Clinical Tuberculosis*, pp. 171-189.

Edlin , B. R., Tokars , J. I., Grieco , M. H., Crawford , J. T., Williams , J., Sordillo , E. M., Ong , K. R., Kilburn , J. O., Dooley , S. W., Castro , K. G., Jarvis , W. R. and Holmberg , S. D. (1992) An Outbreak of Multidrug-Resistant Tuberculosis among Hospitalized Patients with the Acquired Immunodeficiency Syndrome. *New England Journal of Medicine, 326*(23), pp. 1514-1521.

Espinal, M. A. (2003) The global situation of MDR-TB. *Tuberculosis, 83*(1), pp. 44-51.

Feng, Z. and Barletta, R. G. (2003) Roles of Mycobacterium smegmatis d-Alanine:d-Alanine Ligase and d-Alanine Racemase in the Mechanisms of Action of and Resistance to the Peptidoglycan Inhibitor d-Cycloserine. *Antimicrobial Agents and Chemotherapy, 47*(1), pp. 283-291.

Fischl, M. A., Uttamchandani, R. B., Daikos, G. L. and et al. (1992) AN outbreak of tuberculosis caused by multiple-drug-resistant tubercle bacilli among patients with hiv infection. *Annals of Internal Medicine, 117*(3), pp. 177-183.

Fu, X., Duc, L. T., Fontana, S., Bong, B. B., Tinjuangjun, P., Sudhakar, D., Twyman, R. M., Christou, P. and Kohli, A. (2000) Linear transgene constructs lacking vector backbone sequences generate low-copy-number transgenic plants with simple integration patterns. *Transgenic Research, 9*(1), pp. 11-19.

Ginsburg, A. S., Grosset, J. H. and Bishai, W. R. (2003) Fluoroquinolones, tuberculosis, and resistance. *The Lancet Infectious Diseases, 3*(7), pp. 432-442.

Herrera, L., Jiménez, S., Valverde, A., Garcý, amp, x, a-Aranda, M. A. and Sáez-Nieto, J. A. (2013) Molecular analysis of rifampicin-resistant Mycobacterium tuberculosis isolated in Spain (1996–2001). Description of new mutations in the rpoB gene and review of the literature. *International Journal of Antimicrobial Agents, 21*(5), pp. 403-408.

Heym, B., Saint-Joanis, B. and Cole, S. T. (1999) The molecular basis of isoniazid resistance in Mycobacterium tuberculosis. *Tubercle and Lung Disease, 79*(4), pp. 267-271.

Johnson, R., Streicher, E., Louw, G., Warren, R., Helden, P. v. and Victor, T. (2006) Drug resistance in Mycobacterium tuberculosis. *Curr Issues Mol Biol, Jul;8*(2), pp. 97-111.

Lee, A. S. G., Teo, A. S. M. and Wong, S.-Y. (2001) Novel Mutations in ndh in Isoniazid-Resistant Mycobacterium tuberculosisIsolates. *Antimicrobial Agents and Chemotherapy, 45*(7), pp. 2157-2159.

Lee, H. Y., Myoung, H. J., Bang, H. E., Bai, G. H., Kim, S. J., Kim, J. D. and Cho, S. N. (2002) Mutations in the embB Locus among Korean Clinical Isolates of Mycobacterium tuberculosis Resistant to Ethambutol. *Yonsei Med J, 43*(1), pp. 59-64.

Mathew, M., Anupam, D. s., Rajnish, M., Abhinav, S. and Khanna, J. (2014) Tuberculosis the great mimicker: 18F-fludeoxyglucose positron emission tomography/computed tomography in a case of atypical spinal tuberculosis. *Tuberculosis spine: Therapeutically refractory disease*, pp. 99-101.

Mdluli, K., Slayden, R. A., Zhu, Y., Ramaswamy, S., Pan, X., Mead, D., Crane, D. D., Musser, J. M. and Barry, C. E. (1998) Inhibition of a Mycobacterium tuberculosis â-Ketoacyl ACP Synthase by Isoniazid. *Science, 280*(5369), pp. 1607-1610.

Middlebrook, G. (1954) Isoniazid-resistance and catalase activity of tubercle bacilli; a preliminary report. *Am. Rev. Tuberc, 69*, pp. 471-472.

Miesel, L., Weisbrod, T. R., Marcinkeviciene, J. A., Bittman, R. and Jacobs, W. R. (1998) NADH Dehydrogenase Defects Confer Isoniazid Resistance and Conditional Lethality in Mycobacterium smegmatis. *Journal of Bacteriology, 180*(9), pp. 2459-2467.

Mokrousov, I., Narvskaya, O., Limeschenko, E., Otten, T. and Vyshnevskiy, B. (2002) Detection of Ethambutol-Resistant Mycobacterium tuberculosis Strains by Multiplex Allele-Specific PCR Assay Targeting embB306 Mutations. *Journal of Clinical Microbiology, 40*(5), pp. 1617-1620.

Morlock, G. P., Metchock, B., Sikes, D., Crawford, J. T. and Cooksey, R. C. (2003) ethA, inhA, and katG Loci of Ethionamide-Resistant Clinical Mycobacterium tuberculosis Isolates. *Antimicrobial Agents and Chemotherapy, 47*(12), pp. 3799-3805.

Ramana Rao, M. V., Parameswari, C., Sripriya, R. and Veluthambi, K. (2011) Transgene stacking and marker elimination in transgenic rice by sequential Agrobacterium-mediated co-transformation with the same selectable marker gene. *Plant Cell Reports, 30*(7), pp. 1241-1252.

Ramaswamy, S. and Musser, J. M. (1998) Molecular genetic basis of antimicrobial agent resistance inMycobacterium tuberculosis: 1998 update. *Tubercle and Lung Disease, 79*(1), pp. 3-29.

Ramaswamy, S. V., Amin, A. G., Göksel, S., Stager, C. E., Dou, S.-J., El Sahly, H., Moghazeh, S. L., Kreiswirth, B. N. and Musser, J. M. (2000) Molecular Genetic Analysis of Nucleotide Polymorphisms Associated with Ethambutol Resistance in Human Isolates of Mycobacterium tuberculosis. *Antimicrobial Agents and Chemotherapy, 44*(2), pp. 326-336.

Rattan, A., Kalia, A. and Ahmad, N. (1998) Multidrug-resistant Mycobacterium tuberculosis: molecular perspectives. *Emerging Infectious Diseases, 4*(2), pp. 195-209.

Rozwarski, D. A., Grant, G. A., Barton, D. H. R., Jacobs, W. R. and Sacchettini, J. C. (1998) Modification of the NADH of the Isoniazid Target (InhA) from Mycobacterium tuberculosis. *Science, 279*(5347), pp. 98-102.

Ryan, K. J., Ray, C. G. and Sherris, J. C. (2004) *Sherris medical microbiology: an introduction to infectious diseases,* McGraw-Hill.

Scorpio, A., Lindholm-Levy, P., Heifets, L., Gilman, R., Siddiqi, S., Cynamon, M. and Zhang, Y. (1997) Characterization of pncA mutations in pyrazinamide-resistant

Mycobacterium tuberculosis. *Antimicrobial Agents and Chemotherapy, 41*(3), pp. 540-3.

Scorpio, A. and Zhang, Y. (1996) Mutations in pncA, a gene encoding pyrazinamidase/nicotinamidase, cause resistance to the antituberculous drug pyrazinamide in tubercle bacillus. *Nat Med, 2*(6), pp. 662-667.

Sherman, D. R., Mdluli, K., Hickey, M. J., Arain, T. M., Morris, S. L., Barry, C. E. and Stover, C. K. (1996) Compensatory ahpC Gene Expression in Isoniazid-Resistant Mycobacterium tuberculosis. *Science, 272*(5268), pp. 1641-1643.

Slayden, R. A. and Barry Iii, C. E. (2000) The genetics and biochemistry of isoniazid resistance in Mycobacterium tuberculosis. *Microbes and Infection, 2*(6), pp. 659-669.

Somoskovi, A., Parsons, L. M. and Salfinger, M. (2001) The molecular basis of resistance to isoniazid, rifampin, and pyrazinamide in Mycobacterium tuberculosis. *Respiratory Research, 2*(3), pp. 164-168.

Sreevatsan, S., Pan, X., Zhang, Y., Kreiswirth, B. N. and Musser, J. M. (1997a) Mutations associated with pyrazinamide resistance in pncA of Mycobacterium tuberculosis complex organisms. *Antimicrobial Agents and Chemotherapy, 41*(3), pp. 636-640.

Sreevatsan, S., Stockbauer, K. E., Pan, X., Kreiswirth, B. N., Moghazeh, S. L., Jacobs, W. R., Telenti, A. and Musser, J. M. (1997b) Ethambutol resistance in Mycobacterium tuberculosis: critical role of embB mutations. *Antimicrobial Agents and Chemotherapy,*

41(8), pp. 1677-1681.

Suzuki, Y., Katsukawa, C., Tamaru, A., Abe, C., Makino, M., Mizuguchi, Y. and Taniguchi, H. (1998) Detection of Kanamycin-ResistantMycobacterium tuberculosis by Identifying Mutations in the 16S rRNA Gene. *Journal of Clinical Microbiology, 36*(5), pp. 1220-1225.

Takayama, K. and Kilburn, J. O. (1989) Inhibition of synthesis of arabinogalactan by ethambutol in Mycobacterium smegmatis. *Antimicrobial Agents and Chemotherapy, 33*(9), pp. 1493-1499.

Takiff, H. E., Salazar, L., Guerrero, C., Philipp, W., Huang, W. M., Kreiswirth, B., Cole, S. T., Jacobs, W. R. and Telenti, A. (1994) Cloning and nucleotide sequence of Mycobacterium tuberculosis gyrA and gyrB genes and detection of quinolone resistance mutations. *Antimicrobial Agents and Chemotherapy, 38*(4), pp. 773-780.

Telenti, A., Philipp, W. J., Sreevatsan, S., Bernasconi, C., Stockbauer, K. E., Wieles, B., Musser, J. M. and Jacobs, W. R. (1997) The emb operon, a gene cluster of Mycobacterium tuberculosis involved in resistance to ethambutol. *Nat Med, 3*(5), pp. 567-570.

Victor, T. C., van Helden, P. D. and Warren, R. (2002) Prediction of Drug Resistance in M. tuberculosis: Molecular Mechanisms, Tools, and Applications. *IUBMB Life, 53*(4-5), pp. 231-237.

World Health Organization (2001) GUIDELINES FOR DRUG SUSCEPTIBILITY

TESTING

FOR SECOND-LINE ANTI-TUBERCULOSIS DRUGS FOR DOTS-PLUS. in.

World Health Organization (2016a) *Tuberculosis*, Available: http://www.who.int/mediacentre/factsheets/fs104/en/.

World Health Organization (2016b) *WHO Global tuberculosis report 2016*.

Xu, C., Kreiswirth, B. N., Sreevatsan, S., Musser, J. M. and Drlica, K. (1996) Fluoroquinolone Resistance Associated with Specific Gyrase Mutations in Clinical Isolates of Multidrug-Resistant Mycobacterium tuberculosis. *Journal of Infectious Diseases, 174*(5), pp. 1127-1130.

Xu, D., Duan, X., Wang, B., Hong, B., Ho, T. H. D. and Wu, R. (1996) Expression of a Late Embryogenesis Abundant Protein Gene, HVA1, from Barley Confers Tolerance to Water Deficit and Salt Stress in Transgenic Rice. *Plant Physiology, 110*(1), pp. 249-257.

Zhang, Y., Garbe, T. and Young, D. (1993) Transformation with katG restores isoniazid-sensitivity in Mycobacterium tuberculosis isolates resistant to a range of drug concentrations. *Molecular Microbiology, 8*(3), pp. 521-524.

Zhang, Y., Heym, B., Allen, B., Young, D. and Cole, S. (1992) The catalase[mdash]peroxidase gene and isoniazid resistance of Mycobacterium tuberculosis. *Nature, 358*(6387), pp. 591-593.

Zimhony, O., Vilchèze, C. and Jacobs, W. R. (2004) Characterization of Mycobacterium smegmatis Expressing the Mycobacterium tuberculosis Fatty Acid Synthase I (fas1) Gene. *Journal of Bacteriology, 186*(13), pp. 4051-4055.